On a H

Dona Herweck Rice

Publishing Credits

Rachelle Cracchiolo, M.S.Ed., *Publisher*
Conni Medina, M.A.Ed., *Managing Editor*
Nika Fabienke, Ed.D., *Content Director*
Véronique Bos, *Creative Director*
Shaun N. Bernadou, *Art Director*
Valerie Morales, *Associate Editor*
John Leach, *Assistant Editor*
Courtney Roberson, *Senior Graphic Designer*

Image Credits: All images from iStock and/or Shutterstock.

Library of Congress Cataloging-in-Publication Data

Names: Rice, Dona, author.
Title: On a hot day / Dona Herweck Rice.
Description: Huntington Beach, CA : Teacher Created Materials, [2019] | Audience: K to grade 3. |
Identifiers: LCCN 2018029721 (print) | LCCN 2018029830 (ebook) | ISBN 9781493899098 | ISBN 9781493898350
Subjects: LCSH: Heat--Juvenile literature. | Temperature--Juvenile literature. | Weather--Juvenile literature. | Vocabulary.
Classification: LCC QC981.3 (ebook) | LCC QC981.3 .R4957 2019 (print) | DDC 551.5/25--dc23
LC record available at https://lccn.loc.gov/2018029721

Teacher Created Materials

5301 Oceanus Drive
Huntington Beach, CA 92649-1030
www.tcmpub.com

ISBN 978-1-4938-9835-0

© 2019 Teacher Created Materials, Inc.
Printed in China
Nordica.082018.CA21800936

Who can use some  on this day?

water

Who can use some

 on this day?

lemonade

Who can use some on this day?

sunglasses

Who can use some  on this day?

sunscreen

Who can use some  on this day?

ice cream

Who can use some  on this day?

ice pops

Who can use some  on this day?

hats

Who can use some  on this day?

umbrellas

Who can use some  on this day?

sprinklers

Who can use some  on this day?

shade

High-Frequency Words

New Words

can day

some use

who

Review Words

on this